WAVES

WAVES
THE ELECTROMAGNETIC UNIVERSE

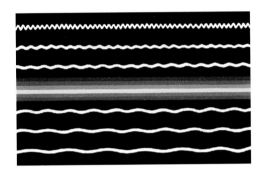

BY GLORIA SKURZYNSKI
Winner of the American Institute of Physics
Science Writing Award

NATIONAL
GEOGRAPHIC
SOCIETY
Washington, D.C.

Distributed by Publishers Group West

NONFICTION BY
GLORIA SKURZYNSKI

ALMOST THE REAL THING
Simulation in Your High-Tech World

GET THE MESSAGE
Telecommunications in Your High-Tech World

HERE COMES THE MAIL

KNOW THE SCORE
Video Games in Your High-Tech World

ROBOTS
Your High-Tech World

ZERO GRAVITY

Copyright 1996 © Gloria Skurzynski

All rights reserved.
Reproduction of the whole or any part of the contents without written permission from the publisher is prohibited.

PUBLISHED BY THE NATIONAL GEOGRAPHIC SOCIETY
1145 17TH STREET N.W.
WASHINGTON, D.C., 20036

Reg Murphy, President and Chief Executive Officer
Gilbert M. Grosvenor, Chairman of the Board
Nina D. Hoffman, Senior Vice President
Will Gray, Vice President and Director of the Book Division
Barbara Lalicki, Director of Children's Publishing
Barbara Brownell, Senior Editor—Mark A. Caraluzzi, Marketing Manager
Staff for this book:
Suez Kehl, Art Director—Greta Arnold, Illustrations Editor—Meredith Wilcox, Illustrations Assistant
Vincent P. Ryan, Manufacturing Manager—Lewis R. Bassford, Production Manager
Corinne Szabo, Picture Researcher—Rick Davis, Indexer—Dale Herring, Editorial Assistant
Design by Rutt Design

Library of Congress Cataloguing in Publication Data

Skurzynski, Gloria.
　Waves : the invisible universe / by Gloria Skurzynski.
　　p.　cm.
　　Summary: Examines diffeerent kinds of electromagnetic waves-
including radio waves, microwaves, light, X rays and gamma rays.
　　ISBN 0-7922-3520-7
　　1. Radiation--Juvenile literature.　2. Electromagnetic waves-
-Juvenile literature.　(1. Electromagnetic waves.　2. Radiation.)
I. Title.
QC475.25.S58　1996
539.2--dc20
　　　　　　　　　　　　　　　　　　　　　　　　　96-11976
　　　　　　　　　　　　　　　　　　　　　　　　　　CIP

Distributed by Publishers Group West
P.O. Box 8843/Emeryville, California 94662
Toll free 1-800-788-3123

This book is dedicated, with great admiration,
to NASA Chief Scientist Dr. France Anne Córdova

The author is grateful to the following people
who so generously shared information:

Jeff Cargill, Sea World of Florida
Dr. France Anne Córdova, NASA Chief Scientist
Dr. Imke de Pater, Deptartment of Astronomy, University of California, Berkeley
Dr. Carl E. Fichtel, Laboratory for High Energy Astrophysics,
NASA-Goddard Space Flight Center
David Finley, National Radio Astronomy Observatory
Dr. G. Randall Gladstone, Department of Space Science,
Southwest Research Institute
Dr. Isabel Hawkins, UC Berkeley Center for EUV Astrophysics
Dr. Mark Hereld, SPIREX Team, University of Chicago
Dr. William S. Lewis, Department of Space Science, Southwest Research Institute
Dr. Eugene Loh, Fly's Eye Cosmic Ray Detector, University of Utah
Steve Maran, NASA-Goddard Space Flight Center
Dr. Daniel Mattis, University of Utah
Dr. Robert Petre, ROSAT Project, NASA-Goddard Space Flight Center
Dr. Scott Severson, SPIREX Team, University of Chicago
Ray Villard, Space Telescope Science Institute
Dr. Mike Walsh, Sea World of Florida
Dr. J. Hunter Waite, Department of Space Science,
Southwest Research Institute

It's fun to play in the ocean. It can be a bit scary, too. You hear the crash of the surf. You smell the salty water, and taste it when it splashes into your mouth. You feel the ocean waves push you, sometimes gently, sometimes hard enough to tumble you head over heels. One after another the waves break against you, and as far out as you can see, more waves keep coming.

There are other kinds of waves. Every second of your life, these other waves surround you, no matter where you are. They strike you, warm you, burn you, and shoot right through your body. Yet you never even notice them. That's because they're invisible—at least most of them are.

They're the waves that make up the electromagnetic spectrum.

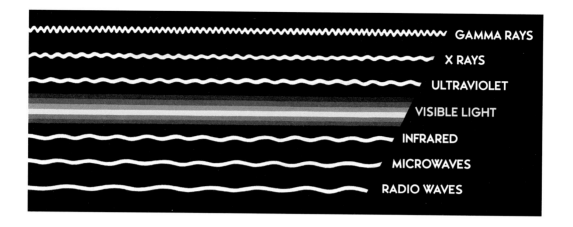

They come from the same kind of electric force that makes your toaster work; and from the same kind of magnetic force that holds your family's shopping list to the refrigerator with a little magnet. When these two forces act together, they make electromagnetic radiation. They make waves.

In some ways, electromagnetic waves are like ocean waves. They have tops called crests, and bottoms called troughs. The distance from one crest to the next (or from one trough to the next) is called a wavelength. Both electromagnetic (EM) waves and ocean waves move energy from place to place.

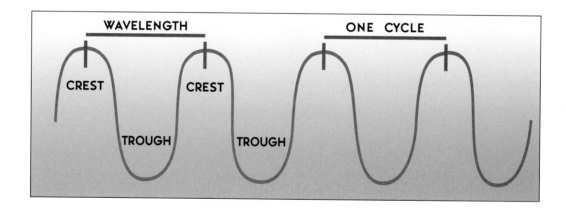

The biggest difference between ocean waves and EM waves is that ocean waves travel across water—the water carries them. EM waves don't need anything to ride on. They can move across emptiness—inside a room, around the earth, or through the vacuum of outer space. In a vacuum, they move at the speed of light.

RADIO WAVES

When Marianne turns on her radio, she's putting part of the electromagnetic spectrum to work. In the whole electromagnetic spectrum, radio waves have the longest wavelengths. From crest to crest, a single radio wave can be miles long. Or it can be as long as a football stadium. Or only as long as a jump rope. Radio waves come in a lot of different lengths, but the ones that carry music and news and baseball games to your AM radio are never shorter than a 12-inch ruler.

Marianne twists the dial looking for her favorite song. She has just changed from one radio frequency to another. "Frequency" means waves per second.

Imagine a long string hanging straight down from the sky to the street outside your bedroom. Now imagine a radio signal sent out from a tower at the edge of town. The first wave passes the string. More and more waves, each with the same wavelength from crest to crest, go past it. A single wave, from one crest to the next or one trough to the next, is called a "cycle." If ten thousand waves pass the string in one second, that means the radio signal has a frequency of ten thousand cycles per second.

Another name for cycle per second is "hertz." In 1888, a 30-year-old German scientist named Heinrich Hertz began to study electromagnetic waves. When he made an electric spark jump between two terminals, he noticed a smaller spark in another,

separate circuit about five feet away. Hertz realized that the energy that caused the second spark must have been a new and different kind of electromagnetic wave. Though they were first called "Hertzian waves," they're now known as radio waves. And in honor of Heinrich Hertz, the scientific term for one cycle per second of electromagnetic energy became "hertz." It's abbreviated "Hz."

Almost always, when a scientist makes a discovery, other scientists build on that discovery to see what can be learned from it and what it can be used for. That happened with radio waves. A 22-year-old Italian scientist named Guglielmo Marconi was the first to understand that if radio waves could cross a room, they should be able to cross an ocean. In 1896, Marconi made the first radio transmitter and sent radio signals for a distance of a mile. Today radios are so small, so powerful, and so inexpensive that almost everyone owns one. Radio waves aren't only used for radio broadcasts; they also carry signals to television sets, cellular telephones, and pagers, or "beepers." They not only cross oceans, they can reach beyond the planets.

It's frequency—the number of cycles to pass a given point in a second—that makes each band of the electromagnetic spectrum different from the other six bands (there are seven in all). The waves with the lowest frequencies have the least amount of energy; those with the highest frequencies have the most energy.

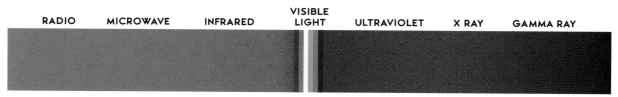

The electromagnetic spectrum extends from radio waves to gamma rays.

MICROWAVES

If radio waves become shorter than 11.8 inches, they get a new name—"microwaves." Microwaves are still radio waves, but they're at the high end of the radio band in the EM spectrum. They can be as short as the thickness of a paper plate, although the ones used in microwave ovens are about five inches long.

When Courtney puts a hot dog into a microwave oven, two billion four hundred and fifty million microwaves per second bombard the meat, bun, and paper plate. The plate stays cool; the bun gets hot; and the frankfurter gets cooked. What's happening in there?

Inside the frankfurter and bun are a lot of water molecules. A molecule is a grouping of atoms. Each water molecule has two atoms of hydrogen and one atom of oxygen. When the microwaves hit them, the water molecules get so excited that they start to dance around—two billion four hundred and fifty million times per second—in perfect rhythm with the microwave frequency. And when the motion of molecules increases, heat increases, too. The heat cooks Courtney's hot dog just the way she likes it.

The paper plate stays cool because it doesn't have any water molecules in it to jump around at the rate of two billion…you know the rest of the number. When numbers get **that** big, scientists and engineers have a shorter way to say them. A million cycles is a megahertz, or MHz. A thousand megahertz, or 1000 MHz, is the same as saying one billion cycles per second. The frequency of the microwaves in Courtney's oven is written as "2450 MHz."

Microwaves do other things than cook food. They're also used for television and communication satellites, and for radar. Radar came first. It was developed by the British in World War II to detect enemy airplanes and ships. A microwave beam would be sent out from an antenna, and when the beam hit something solid like a battleship, it would bounce back, even if the ship was miles away. The reflected waves

work the way a bat's squeak works when it echoes off a cave wall. The bat hears the reflected sound and can tell where the wall is. A radar beam, bouncing back, can show where a battleship or other object is.

After the war, radar tubes powered the first microwave ovens. One was used by a street vendor selling "Radar Dogs." Today, microwaves travel thousands of miles into space to reach communications satellites, which then transmit the waves back to different places on earth. These waves have shorter wavelengths than the ones used in microwave ovens, as short as four-hundredths of an inch. When a wavelength is that short, the frequency is high—three hundred billion cycles per second. Written out, that's 300,000,000,000. It's easier to write 300 GHz, or 300 gigahertz. "Giga" means one billion. If it's hard to remember, you may want to look at this table:

NUMBER	SPELLED	PREFIX	SYMBOL
1,000,000,000,000	one trillion	tera	T
1,000,000,000	one billion	giga	G
1,000,000	one million	mega	M
1,000	one thousand	kilo	k

INFRARED

Stephanie and her brother James are having a fight over the television remote control. They're not really in a big fight; they're just wrestling on the floor while Stephanie tries to grab the controller. They have this tug of war several times a day.

This time Stephanie wins. Aiming the controller toward the TV, she pushes a button to change the channel.

It's necessary for Stephanie to point directly at the television set. The infrared waves that control the set have to travel in a straight line, with nothing in the way to block the waves. This is called "line of sight" transmission.

When she pushes the button, a microchip in the controller starts up a vibration. It produces waves with a frequency in the infrared region of the EM spectrum: 100 billion to 100 trillion cycles per second. That brings up the last of the "for short" prefixes: tera. One terahertz (THz) is one trillion hertz, or cycles per second.

Just as radio waves can carry radio and television signals, these infrared beams can carry codes. Inside the television set, another microchip identifies each code and sends a signal to raise or lower volume, change channels, turn the set on and off, record a program, or do whatever Stephanie wants it to do.

Since the frequencies of infrared beams are so high, the wavelengths are short: from four-hundredths of an inch to only four-hundred thousandths of an inch.

If you stand in sunlight, you feel warmth from the sun's infrared radiation. It's the same kind of infrared radiation that works in Stephanie's remote controller. Your body, too, gives off radiant heat because you're warm-blooded. Infrared radiation from your body can be captured on special film that makes a picture called a thermogram.

Because killer whales are mammals—marine mammals—they have body heat, too, just like you do. At Sea World of Florida, veterinarian Mike Walsh (wearing boots) videotapes Shamu the killer whale at infrared frequencies. With Shamu's mouth wide open, Dr. Walsh can aim the camera at the whale's teeth to tell whether they're healthy. A sore tooth would give off more heat than the others, and would show up on the thermogram at different wavelengths.

Dr. Walsh wears a virtual-reality headset that brings the infrared image right in front of his eyes. That way he can be sure he's looking where he wants to look, without worrying about light reflecting off the water into his eyes, or the whale moving while he takes a quick look at the monitor. Wearing the headset, Dr. Walsh can see in real time the colors and numbers of the thermogram he's taking of Shamu.

A thermogram appears in "false color" rather than the true color of the visible world. Different colors indicate different temperatures. Shamu's skin and the water are both about 12°C (centigrade), or 54°F (fahrenheit). Because they're the same temperature, they're the same color in the thermogram, even though in true, visible light they look different. Shamu's eye is ringed with red because there's more warm blood in an eye area. The mouth is open in a v-shape; it's outlined in yellow. The tongue is outlined in red. A fish being thrown into the whale's open mouth looks dark blue because a cold-blooded fish doesn't register any heat at all.

At the top of the thermogram, right under the color bar, temperatures are listed in centigrade. From 12°C (the water) to 16°C (the eye) doesn't seem like much of a temperature difference, but even small heat changes would point out problems if the whale had an illness or an injury. That's why a thermogram is a valuable tool for checking the health of a creature that's hard to examine close-up.

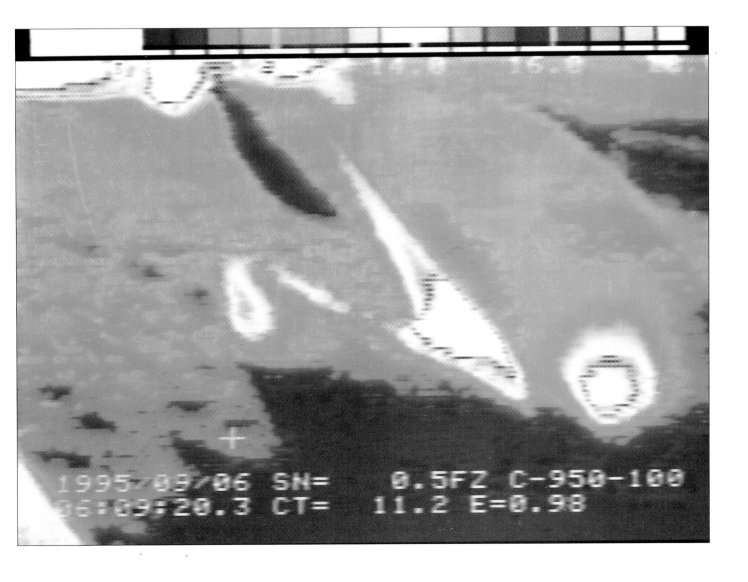

VISIBLE LIGHT

The bands of the electromagnetic spectrum don't have exact starting and stopping places. You can't draw a line and say, "Microwaves end here, and infrared waves begin on this other side of the line." The spectrum is seamless: the waves change smoothly from one band to another.

Just above the infrared band of the electromagnetic spectrum, you reach the part of the spectrum that's most important to human beings. Here the waves become very short—between 30 millionths of an inch and only 14 millionths of an inch. The frequencies become very high—ranging from a hundred trillion to a thousand trillion cycles per second.

And something wonderful happens. You can see the electromagnetic waves!

This is the visible part of the EM spectrum, the only part that your eyes are built to respond to. Not only your eyes, but the eyes of every creature on earth—the two eyes of a mammal or a bird, the four eyes of an insect, the six or eight eyes of a spider, the hundred-some eyes of an ocean-dwelling scallop have evolved to be able to see these particular light waves.

This happens even though the visible light band is just a small part of the whole electromagnetic spectrum.

Paul and Tom are watching their grandfather mix paint. As he stirs pigment into white paint, it brightens with color. When you were little, in preschool or

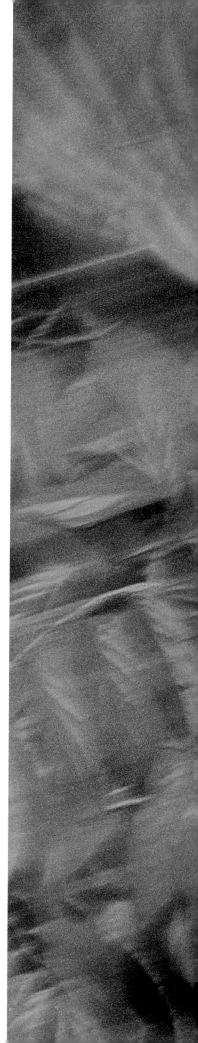

A macaw in motion is a dazzling example of the visible spectrum.

kindergarten, you too learned how to mix colors. Using poster paints, you made green by adding yellow to blue, or orange by mixing red with yellow, or purple by mixing red and blue.

Your teacher told you that red, yellow, and blue were the primary colors, and that's true—for paints. But light—light's different. Light has energy!

With visible light, the primary colors are red, green, and blue. On your television screen, tiny dots of these 3 colors can blend into 16 million different color combinations of light.

To see the pure colors of the visible spectrum, you need a prism. While his

friend Eric watches, Briggs holds a prism in a beam of sunlight. The beam passes through the prism, is bent, and spreads out into the six colors of the spectrum: red, orange, yellow, green, blue, and violet. Some people say there's an extra color, indigo, that falls between blue and violet. Each of these colors has its own wavelength and frequency, but the edges do blend.

Isaac Newton, a brilliant scientist who lived in the seventeenth century, was the first to discover this. He covered a window and cut a small round hole in the covering to let in a ray of sunlight. When he passed the light through a prism—exactly as Briggs and Eric are doing—the light spread out into the colors of the rainbow. But this spectrum wasn't round, like the hole in the window blind. It was a wide band of color, and each color appeared separate and distinct. Newton then recaptured the spread-out band of colors in another prism, and found that they combined into a round spot of white light. Until that time, scientists had believed that colors were a mixture of black and white in varying proportions.

After his experiment, Newton came to believe that colors were formed by tiny particles, and that different colors came from different-size particles of light. He was not quite right about that (check page 22). Even brilliant scientists can miss seeing the whole truth sometimes! But Newton's experiments with prisms did prove that white light is a combination of all the colors of the rainbow.

Late on a summer afternoon, Danny stands with his back to the sun. By spraying a fine mist of water from the garden hose, he creates his own rainbow. The water drops act like tiny prisms, separating white light into the bands of the visible spectrum. In the picture below you'll see a rainbow that nature has made. It's much bigger than Danny's, but it looks the same. The sun's rays strike raindrops falling from clouds overhead, bend the light, and spread it apart into the colors of the spectrum.

Notice that the colors appear in exactly the same order—in the rainbow in the sky; in Danny's backyard rainbow; and in the spectrum Briggs created with his prism: always red, orange, yellow, green, blue, and violet. Starting with red, each color that follows has a shorter wavelength and a higher frequency than the one before it.

A rainbow in the sky only happens in the early morning or the late afternoon, when the sun is no higher than 42 degrees above the horizon.

At any time of day, no matter where the sun is, Jamal can make what look like rainbows appear in bubbles. But they're not really rainbows.

The colors of the rainbow come from light waves that move out in straight lines. They stay separated. The colors in Jamal's bubbles come from light waves

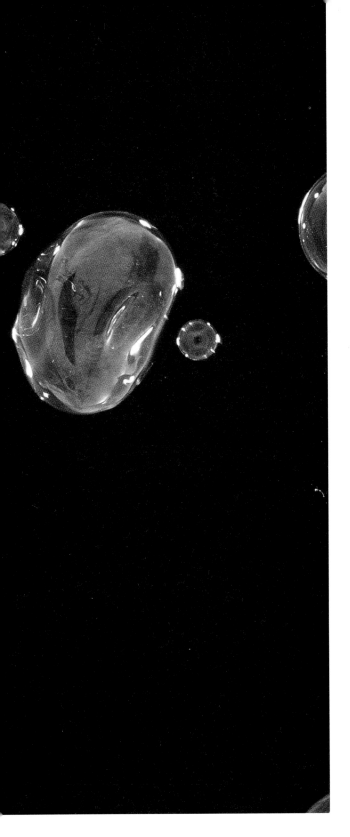

that cross each other. That's because bubbles are thin and transparent. Light hits the outside surface of the thin film of a bubble, and reflects off it. But light also goes through the film, and bounces off the underneath surface. Since the bubble film is a little thicker in some places and thinner in others, different wavelengths of light pass through one another, strengthening some colors and destroying some. This mixing up of wavelengths is called "interference." It causes the colors to swirl around in no particular order.

All the colors of the visible spectrum combine in white light to illuminate everything you see. If there's no light, you can't see color—just try picking out some clothes from your closet in the dark. To see, you have to have light, from the sun or from an electric bulb or a fire or some other source of radiation. Everything you look at has its own distinct color—red or pink or chartreuse or aqua or any of the 16 million other combinations that are possible in the visible world when light waves combine.

Each object you see—your dog, your basketball, your classroom, your teacher—is made up of molecules. When light shines on them, the molecules on the surface absorb most of the wavelengths in the visible spectrum. When they **don't** absorb a certain color's wavelength, it's reflected.

The colors you see are not the ones that are absorbed, but the ones that are reflected. Why is a tomato red? Because the molecules in the tomato absorb the wavelengths of orange, yellow, green, blue, and violet. They reflect the wavelength of red, so to your eyes, the tomato looks red.

Chlorophyll in leaves absorbs wavelengths in the violet, orange, and red portions of the visible spectrum, and reflects wavelengths in the blue, green, and yellow portions of the spectrum that combine to make green. So just about every growing plant on earth looks green, the color that its leaves reflect. Why does this happen? Scientists aren't really sure. It's one of the mysteries about light and the EM spectrum that hasn't yet been completely solved.

Scientists are always working hard to find the answers to such mysteries, and to explain them beyond any doubt. Three centuries ago when light was first studied, men of science (all scientists were men back then) had different opinions about whether light rays were waves, or thin streams of tiny particles. Isaac Newton believed in the particle theory; a Dutch physicist named Christiaan Huygens said that light was a wave. Later scientists examined, analyzed, experimented, and debated the "wave theory" versus the "particle theory." It wasn't until this century that they agreed that light rays are probably both waves **and** particles.

Even though electromagnetic radiation can be identified, produced, and made to operate all sorts of things, it still isn't completely understood. Engineers who create products are not always sure why every part of them works, because all the scientific reasons haven't yet been discovered.

That's what makes being a scientist so exciting—there are so many questions about the physical world that are still unanswered. It's such an enormous puzzle that to discover even one small piece of it makes a lifetime of study worthwhile. But luckily for you and for everyone else, you don't have to know exactly **how** something works to **make** it work. When prehistoric people discovered fire, they may not have known how to explain it, but they quickly figured out that they could use it to keep warm.

ULTRAVIOLET WAVES

Ultraviolet waves won't keep you warm. Danny is kneeling in ultraviolet waves that are shorter than visible light waves. UV waves start around 15 millionths of an inch in length.

Higher up on the electromagnetic spectrum than visible light (meaning they have a higher frequency: more than one thousand trillion cycles per second), they can't be seen by Danny's eyes. But when UV (ultraviolet) waves hit certain materials that "fluoresce"—glow—they produce visible light.

Special "black light" fluorescent tubes send out energy in ultraviolet wavelengths only a billionth of an inch long.

These tubes are different from the "white" fluorescent lights in the hallways of buildings. Black lights are made of a special kind of glass that lets the ultraviolet waves pass through.

Danny's white shirt, pants, and the card he's holding fluoresce, but his hair, skin, belt, and the letters on the card that spell "HI" don't fluoresce. So in the darkened room, they're hard to see.

Ultraviolet radiation isn't always so much fun. Kathy's sunburn is already starting to hurt, and she hasn't even finished swimming. On a bright sunny day, the ultraviolet part of the sun's radiant light burns skin—the longer Kathy is exposed to UV waves, the more damaged her skin will become.

A certain amount of ultraviolet radiation is good for you. It allows your body to produce vitamin D, and that makes your bones strong. But too many sunburns over a lifetime can cause serious problems for skin. And not just in humans.

In many parts of the world, frogs, toads, and salamanders are disappearing. This worries scientists—after all, frogs and toads have been on earth since before the dinosaurs. Biologists are searching for explanations to this mystery. Possible answers are pollution, destruction of habitat, and the thinning of the ozone layer. The ozone layer shields earth from the sun's harmful ultraviolet rays. But when the ozone layer grows thinner, more UV radiation can reach earth. Since some of the disappearing creatures lay eggs in shallow water where sunlight shines directly on them, biologists wonder if too much UV radiation may be killing the eggs. It hasn't yet been proven, but certain experiments point to this possibility.

X RAYS

The higher you go in the electromagnetic spectrum, the more powerful the rays become. The next band above UV waves is X rays. Notice that the name has changed from "waves" to "rays." From now on, as wavelengths grow infinitesimally small, and frequencies become almost uncountably high, EM radiations will be called "rays." And more often now you'll hear about "particles"—little packets of electromagnetic energy. Another term for these packets of EM radiation is "photons."

X rays have wavelengths around one ten-billionth of an inch or more. Their frequencies are about one million trillion cycles per second, and their energy is so great that just a brief burst can kill diseased cells all the way **inside** your body. Ultraviolet waves damage the skin on your body's surface, but X rays go much deeper.

If you go rollerblading, it's always a good idea to protect yourself with knee pads, elbow pads, and a helmet. Sometimes, though, even with protection, you can fall and break your bones. When Adam fell, he broke both the bones of his right wrist. In the hospital a radiation technician took an X ray of Adam's broken bones.

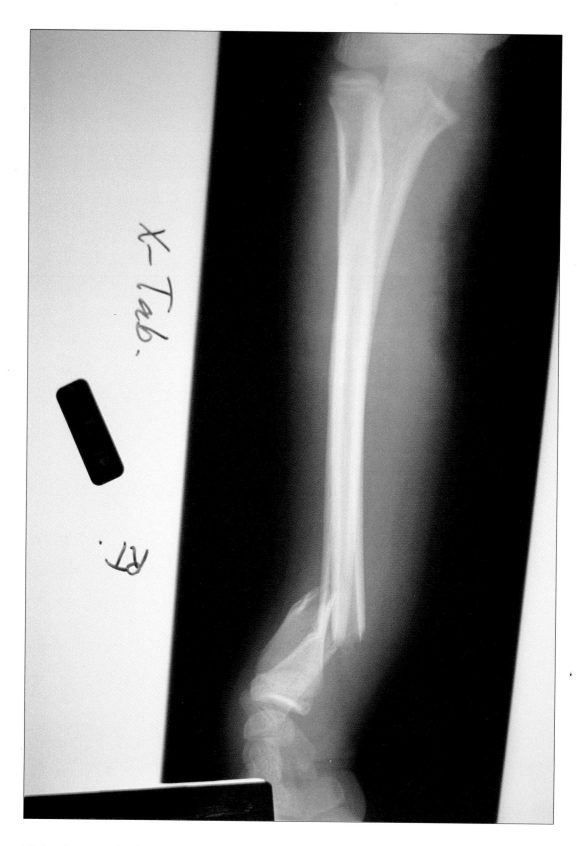

High voltage sends electrons across a vacuum tube to produce X rays.
Bones absorb some radiation to create shadows on film.

When X rays were discovered only a little more than a hundred years ago, newspapers and journals were filled with the story. In the first year alone, they published more than a thousand articles on the mysterious rays. Never before had a scientific discovery caused such widespread excitement. Now there was a way to see inside people without cutting them open!

The discoverer of X rays, Wilhelm Konrad Roentgen, had made an X-ray image of the bones in his wife's hand, which she held on a photographic plate for 15 minutes while Roentgen worked the X-ray tube. In 1901, for his discovery, Roentgen received the very first Nobel Prize for Physics. In 1923 he died of cancer.

Another Nobel Prize winner-to-be, the famous Polish-born scientist Marie Curie, equipped a car with X-ray equipment and took it to the battlefields of France during the First World War. With the help of her 17-year-old daughter Irene, she X-rayed soldiers wounded by bullets and shrapnel. X rays were still new enough that the soldiers were a little afraid of them. "Will it hurt?" one of them asked. "No more than having your picture taken," Marie Curie answered.

Neither Marie nor Irene Curie realized how much the X rays were hurting—them! Their only protection against radiation was the goggles and white gloves they wore, which were really no protection at all. Marie, who received even heavier doses of radiation in her Paris laboratory, eventually died of anemia. Irene later died of leukemia caused by handling the X-ray equipment on the battlefield.

Slowly, scientists began to understand that too much exposure to X rays was deadly. That's why, when Adam's wrist was X-rayed, the technician stood behind a shield made with a layer of lead. X rays can't penetrate lead.

With her husband, Pierre Curie, Marie Sklodowska Curie discovered radium.

GAMMA RAYS

The next step higher in the electromagnetic spectrum—and the last part of the known spectrum—is the gamma-ray band. Gamma rays are extremely high energy. They are no wider than the nucleus of an atom, and have a frequency around one hundred million trillion cycles per second—some higher, some lower. Gamma rays are as much as ten billion times more energetic than visible light. These powerful rays get released into earth's atmosphere when a nuclear bomb explodes, and are given off by the radioactive fallout. You definitely don't want to be anywhere near them.

If a gamma ray passes through a healthy human cell, it can knock electrons from some of the cell's atoms. After enough of this kind of damage, the cell may die. That's why nuclear accidents cause radiation sickness.

Yet these same gamma rays, used cautiously, can destroy diseased cells and make patients healthy again. Gamma rays, and the X rays used in radiation therapy, are photons. But one gamma-ray photon can carry a thousand times more energy than an X-ray photon. They may be small, but they're powerful!

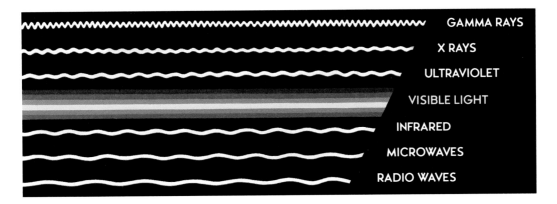

Every part of the EM spectrum you've seen up till now, from radio waves to gamma waves, has been put to work on earth. During the last hundred years, scientists and engineers have discovered countless ways to create and use electromagnetic energy. From the time you get up in the morning until you go to bed at night, your life is filled with EM waves that light your way, cook your food, entertain you, and help you pick out the color shirt you want to wear. Go outside on a sunny day and you drown in an ocean of EM waves! Radio waves, radar from police cars looking for speeders on the highway, waves from garage-door openers, cellular phone calls zapping past you at the speed of light, TV broadcasts—all of them wash in, around, and over you. And over everyone else on earth.

But far beyond earth, the skies are filled with naturally occurring electromagnetic waves that you never know about. If you could see gamma rays, you'd notice bright flares of them in the sky, as dazzling as an entire Fourth-of-July's worth of fireworks. They're called gamma-ray "bursts," or "bursters," and they're one more sample of a scientific mystery. They could be coming from nuclear reactions in stars in our own Milky Way galaxy, or they could come from as far as eight billion light years away—more than halfway to the edge of the universe.

Usually once a day, or it can be as often as five times a day, there's a gamma-ray burst. It may last a few thousandths of a second, or as long as a few minutes. The energy it releases in just a few seconds may be as much as our sun puts out in a thousand years!

If bursters are that powerful, why can't you see them? First, because gamma rays aren't part of the visible spectrum. Second, because most electromagnetic waves from outer space don't travel through earth's atmosphere. Some of them do—radio waves, certain infrared waves, all visible light and part of the ultraviolet band of the EM spectrum (the part that gives you sunburns). But earth's atmosphere blocks many infrared waves, most ultraviolet waves, all X rays, and all gamma rays so they can't reach you. That's a good thing, because if UV waves can burn your skin, think what X rays or gamma rays would do to you if they shone on the earth's surface!

Suppose, suddenly, you could see things at every wavelength. What if you were looking at an airplane—a jumbo jet passenger plane—flying through the sky above you. As usual, in the visible-light wavelengths, you could see the silvery skin of the big aircraft, and the contrails it leaves behind.

This image of the high-energy gamma-ray sky came from data gathered by the Compton Gamma Ray Observatory.

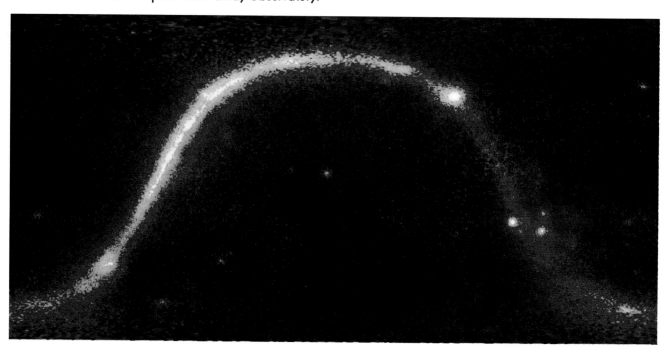

Radio and radar signals travel in invisible waves between airplanes and control towers.

But what if the radio signals the pilot sent to the airport tower suddenly became visible? And you could begin to see the infrared waves given off by the hot jet engines. And the sun's ultraviolet rays made the plane look purple, and your X-ray vision let you see flaws in the metal shell of the plane, and you noticed a gamma-ray burster exploding right above the wings.... You'd grow dizzy from all the images bombarding you. But you'd also discover a lot more about that airplane than you could tell by visible light alone.

WHAT THE WAVES TELL US

That's why scientists now study the whole universe at many different wavelengths: because stars and galaxies and all the other objects in the sky send out radiation at all wavelengths. Special telescopes and cameras examine the skies in radio wavelengths, and in infrared, visible light, ultraviolet, X-ray, and gamma-ray wavelengths. Some of these telescopes are on earth, while others orbit the earth, high above the atmosphere that blocks so many of the high-frequency, short-length waves.

In 1994, the most spectacular astronomical event of the century took place, and just about every telescope on earth and in space, at every wavelength, watched it happen.

It all started on March 23, 1993. Three astronomers—David Levy, and the husband-wife team of Eugene and Carolyn Shoemaker—were taking pictures of the sky at Mount Palomar Observatory in California. Two nights later, when

Carolyn Shoemaker had developed the film and was examining the pictures, she suddenly sat up straight in her chair. "I don't know what this is, but it looks like a squashed comet," she exclaimed.

As the whole world soon learned, the comet wasn't squashed, it was in pieces: 21 pieces, strung out behind each other like a pearl necklace. And all those pieces, one after the other, were going to smash into the planet Jupiter at speeds up to 130,000 miles per hour. The collision happened in July of 1994.

When the comet fragments, named for letters of the alphabet, began to hit Jupiter, astronomers got busy capturing images at five separate bands of the electromagnetic spectrum. You might expect the images to look pretty much alike since they're all showing the same thing—pieces of comet crashing into a big planet. On the next few pages, you'll see just how different those various wavelength images turned out to be.

VERY LARGE ARRAY

On the desert sands of New Mexico, 27 radio telescopes line up in the shape of a Y. The telescopes look like dishes tilting toward the heavens, but very **big** dishes—each one is 82 feet wide and 96 feet high. Together, they're called the Very Large Array, and they search the universe for radio waves.

Even before 1994 the Very Large Array had collected radio-wave images of signals emitted by energetic electrons in Jupiter's powerful radiation belts. Most of the early images were made from wavelengths somewhere between Marianne's radio (p. 9) and Courtney's microwave oven (p. 11).

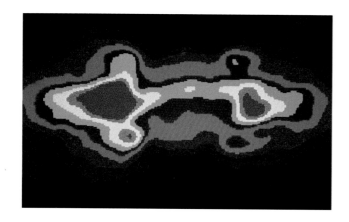

The radio astronomers were eager to make images of the crash, but they expected the impacts to raise so much dust that Jupiter's radio signals would grow dim. Just the opposite happened. The picture that looks like a brightly colored salamander was made right after the collisions. It shows a big **increase** in the radio signals!

SPIREX

At the South Pole, in the middle of snow and ice and glaciers, there's an infrared telescope called SPIREX (for South Pole Infrared Explorer). In July 1994, when the South Pole was dark 24 hours each day, physicist Hien Nguyen focused SPIREX on the planet Jupiter. SPIREX took pictures in infrared at wavelengths about halfway between Stephanie's television remote controller (p. 13) and Shamu the killer whale's thermograph (p. 15). The biggest, most spectacular image of all was the explosion of Fragment G shown here, but many others were almost as brilliant.

HUBBLE

Then there was the Hubble. The Hubble Space Telescope had disappointed everyone at first, right after it was launched, when it sent back fuzzy pictures from its orbit around earth. But during a thrilling space mission in early December 1993, U.S. astronauts caught and repaired the nearsighted Hubble. While the Space Shuttle and the Space Telescope circled earth together, Astronaut Kathryn Thornton maneuvered Hubble's "eyeglasses" into place. She managed to lift this refrigerator-size piece of equipment because in space, heavy things are almost weightless. Astronaut Tom Akers is working at lower left.

The next time the Hubble sent back pictures of the sky, the images looked much sharper. From 380 miles above earth in July of 1994, the Space Telescope was in position to get a great view of Jupiter's biggest hits.

Everything worked perfectly. In this picture made with the visible-light camera, you can see brown smudges at the bottom of Jupiter. From left to right, they're the impact sites (the places where the pieces hit) of Fragments A, E/F, H and Q1. The smudges stayed there for months after the crash.

EUVE

There's a region higher in the spectrum than the ultraviolet waves that burned Kathy's skin; it's the extreme ultraviolet band. In 1992 a satellite called the Extreme Ultraviolet Explorer, or EUVE for short, reached orbit. EUVE made images of Jupiter before the comet collisions. And then it made images again, after the impacts sent up clouds, dust, and fireballs 930 miles high. Following the crashes, the images grew four times brighter! A black circle has been added here to outline the edges of Jupiter. The sideways circle shows the orbit of Jupiter's moon, Io.

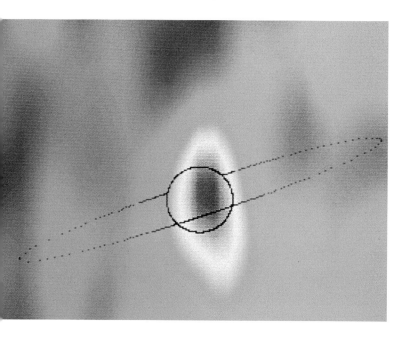

ROSAT

ROSAT is a shortened name: RO for Roentgen, the man who discovered X rays (p. 27), and SAT for satellite. ROSAT explores the skies at X-ray wavelengths. It was launched from Florida in 1990 on the top of a Delta rocket.

When ROSAT took pictures of the comet collision with Jupiter, they looked upside down. After all, the fragments hit at the bottom of Jupiter's disk. But the images ROSAT sent to earth showed a big, bright spot at the **top** of Jupiter, near its north pole.

After a lot of thought, astronomers said that maybe the impact of Fragment K had triggered an electromagnetic disturbance. It traveled through the magnetic field and went off in an X-ray burst in Jupiter's aurora. Like earth, Jupiter has auroras at both the north and south poles. But all this is still a theory. Information collected from space in just a few minutes can take years to analyze.

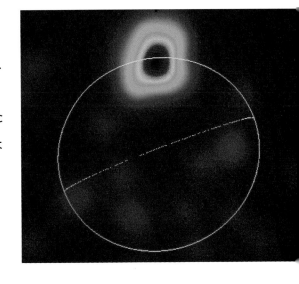

None of these comet-collision images would have been possible 20 years ago. Or even 10 years ago. Only when observatories began orbiting above earth, and computers became powerful enough to sort out their data, could astronomers learn so much about the universe.

Because Hubble Telescope cameras have taken long-exposure pictures of parts of the universe, you can now look far into space and see greater detail, than any humans ever could before. Some of the dim galaxies pictured below are more than ten billion light years away. But you can see them because infrared, visible light, and ultraviolet images, gathered above earth's atmosphere, were combined to give you this true picture.

Using all the different wavelengths of the electromagnetic spectrum, scientists can now search for answers to questions such as:

The Hubble Telescope looks toward the edge of the universe.

A birthplace of stars

How old is the universe? How big is it? Will it expand forever? How are stars and galaxies born?

A magnificent image from the Hubble telescope has given us a partial answer to that last question. It shows columns of cosmic gas and interstellar dust rising six trillion miles from bottom to top. Intense ultraviolet radiation heats those clouds, boiling away the gas. This leaves behind denser pockets of material that, over millions of years, get pulled together by gravity to make new stars.

Some of those stars are likely to have planets. So far just a small number of planets have been detected outside our solar system. No one has actually seen them—not yet! But astronomers know they exist because of the effects they have on the stars they orbit. Planets pull on their mother stars, making them wobble just a bit, just enough to be noticed here on earth.

Of the planets recently discovered, at least two could have the right temperatures to form water. And where there's water, life can exist.

That leads to one enormous question still to be answered, perhaps the most important question of all:

Are we alone in the universe?

You and everyone else can't help wondering if intelligent creatures are out there somewhere, living on those newly discovered planets outside our solar system. Or in a galaxy even farther away.

For more than 35 years, people have listened for messages from outer space. With large radio telescopes located in Australia, Argentina, Puerto Rico, Massachusetts, Ohio, and California, scientists eavesdrop at frequencies from one billion to ten billion cycles per second, at wavelengths from 12 inches to a little over an inch.

They've caught billions of signals, but not the right kind. What the searchers want to hear is a clear, clean, distinct radio signal that repeats itself. One that stands out from the noisy jumble of signals that originate on earth every second—radio and television shows, car phones, beepers, and more.

There are always random radio signals showering down on us from space, too. They stream in from within the Milky Way and from other galaxies farther away. These signals are natural; they come from physical reactions taking place in stars and galaxies. And certain clear, repeated signals do arrive from space, but they're from binary stars, pulsars, and other rotating bodies.

What searchers hope for are signals sent from space—on purpose. Ones created by an unmistakable intelligence.

In Arecibo, Puerto Rico, the world's largest radio telescope listens for signals from outer space.

More than once in the past 35 years, the right kind of signal **may** have been heard. But each time, it was heard only once, and didn't last long enough for radio astronomers to locate the source.

One day, maybe next year or in the next century or 20 centuries from now, a radio signal may reach earth carrying a message. And if you or other earth people can identify the frequency and wavelength the signal travels on, and if you can decipher its message....

Then nothing will ever be the same again.

GLOSSARY

astronomer—a scientist who studies the motion and composition of celestial bodies

black light—the longer-wavelengths of ultraviolet radiation

chlorophyll—the substance in plants that makes them green and helps them convert light to energy

crest—the highest point of a wave

cycle—a wavelength

electromagnetic radiation—waves of electrical and magnetic energy that travel through space at the speed of light

electromagnetic spectrum—the full range of electromagnetic radiation from the longest to the shortest wavelengths and the lowest to the highest frequencies

earth's atmosphere—all the air surrounding Earth

energy—the ability of matter to do work

extreme ultraviolet—the high-energy end of the ultraviolet portion of the electromagnetic spectrum

Extreme Ultraviolet Explorer—an Earth-orbiting observatory with four on-board telescopes that scan the sky at extreme ultraviolet wavelengths

fluoresce—to emit energy as visible light when ultraviolet energy is absorbed

frequency—the number of wave crests or troughs passing a given point in a second

gamma rays—the shortest-wavelength, highest-energy form of electromagnetic radiation

galaxy—a system that contains from millions to billions of stars and cosmic gas and dust

gigahertz—a frequency of one billion cycles per second; abbreviated GHz

hertz—a unit of frequency that equals one cycle per second; abbreviated Hz

Hubble Space Telescope—U.S. astronomical observatory that orbits earth and captures sky images with visible-light and ultraviolet-wavelength cameras

infrared—a band of electromagnetic radiation with a lower frequency and longer wavelength than visible red light

interference—the interaction between two sets of out-of-phase electromagnetic waves that reinforce or neutralize each other

interstellar dust—dust that exists in space between the stars

kilohertz—a frequency of one thousand cycles per second; abbreviated kHz

line of sight—a straight line between the origin of an electronic signal and the object it contacts

megahertz—a frequency of one million cycles per second; abbreviated MHz

microwaves—electromagnetic waves at the high end of the radio-wave band of the EM spectrum, but less energetic than infrared waves

molecule—the smallest particle of a substance that still has the chemical properties of that substance. Molecules are made up of two or more atoms.

ozone layer—the region of earth's atmosphere, about 12 to 30 miles above earth's surface, that prevents most ultraviolet and other high-energy radiation from reaching earth

particle—the smallest component of any kind of matter

photon—a packet of electromagnetic energy; a chargeless particle that travels at the speed of light

primary colors—in paint: red, yellow, and blue; in light: red, green, and blue

prism—a transparent piece of glass or crystal that separates visible light into the colors of the spectrum

radar—a system that uses microwaves to locate distant objects by measuring waves reflected from the objects' surfaces

radiant heat—the heat that radiates from a warm body

radiation—a flow of energy; photons or particles traveling in waves; the process of transmitting energy through space

radioactive fallout—emission of energy and subatomic particles from a nuclear explosion, carried by wind currents

ROSAT—an earth-orbiting observatory that studies the skies at X-ray wavelengths never before imaged from space

spectrum—the entire range of electromagnetic radiation, in bands separated according to frequencies and wavelengths; or, the separation of light into colors

SPIREX—a telescope located in Antarctica that studies the skies at infrared wavelengths

terahertz—a frequency of one trillion cycles per second; abbreviated THz

thermogram—an image showing differences in radiated heat

trough—the lowest point in a wave

ultraviolet—a band of electromagnetic radiation just beyond the frequencies and wavelength of visible light

UV—ultraviolet

vacuum—a space completely empty of matter

Very Large Array—a radio telescope in New Mexico made up of 27 big dishes arranged in the shape of a Y

virtual reality—an environment in which users can interact with computer-generated images

wavelength—the distance from crest to crest or trough to trough of any kind of wave

white light—all the colors of the spectrum combined; the portion of electromagnetic radiation that's visible to human eyes

X rays—electromagnetic radiation with wavelengths and frequencies between ultraviolet radiation and gamma rays; or, images made with X-ray technology

INDEX

Illustrations are indicated by **boldface**. If illustrations are included within a page span, the entire span is boldface.

Akers, Tom **36**, 37

"Black light" fluorescent tubes 23, 44
Bubbles 20-21, **20-21**

Chlorophyll 22, 44
Colors 18-22
   absorption and reflection of 21-22
Curie, Irene 27
Curie, Marie 27, **27**
Cycle **see** Electromagnetic waves

Electromagnetic spectrum 10, 16, 44
   diagrams showing bands of **8, 10, 29**
Electromagnetic waves 7-8, 29-31
   cycles 9
   frequency 9, 10, 44
   wavelength 8, 45
   see also Gamma rays; Infrared waves; Microwaves; Radio waves; Ultraviolet waves; Visible light; X rays
EUVE (Extreme Ultraviolet Explorer) 38, **38**, 44
   image of Jupiter from **38**

Fluorescence 23, **23**, 44
Frequency **see** Electromagnetic waves

Gamma rays 28-31, 44
   "bursters" 29-30
Gigahertz (GHz) 12, 44

Hertz, Heinrich, 9, 10
Hubble Space Telescope **36**, 36-37, 44
   galaxies seen with 40, **40**
   image of Jupiter from **37**
   planets discovered with 41
   star formation seen with 41, **41**
Huygens, Christiaan 22

Infrared waves 13-15, 44
   instruments for detecting 14-15, **14-15**, 35, **35**
Interference, wavelength 21

Jupiter (planet) **33-39**

Kilohertz (kHz) 9, 44

Levy, David 32

Marconi, Guglielmo, 10
Megahertz (MHz) 11, 44
Microwaves 11-12, 44
   ovens **11, 12**

Newton, Isaac 18, 22
Nguyen, Hien 35
Nuclear bomb explosion **28**

Ozone layer, 24, 45

Photons 25, 29, 45
Prisms 18, **18,** 45

Radar 11-12, 45
Radiation 25, 45
   effects of high-frequency 24, **24**, 25, 27, 29
Radio telescopes 34, **34**, 42, **42-43**
   search for extraterrestrial life 42-43
Radio waves 9-10
   instruments for detecting 34, **34**
Radios 9-10, **9-10**
Rainbows **19**, 19-20

Roentgen, Wilhelm Konrad 27
ROSAT 39, **39,** 45
   image of Jupiter from **39**

Satellites 12, **38-40**
Shoemaker, Carolyn 32-33
Shoemaker, Eugene 32
Shoemaker-Levy comet 32-33, **32-33**
   collision with Jupiter **33-39**
SPIREX (South Pole Infrared Explorer) 35, **35**, 45
   image of Jupiter from **35**
Star formation 41, **41**

Terahertz (THz) 13, 45
Thermograms 13-15, 45
   Shamu the killer whale examined with 14-15, **14-15**
Thornton, Kathryn **36**, 36-37

Ultraviolet waves 23-24, 45
   extreme ultraviolet band 38
   instruments for detecting 38, **38**

Very Large Array (radio telescopes) 34, **34**, 45
   image of Jupiter from **34**
Visible light 16, 18-22
   instruments for detecting **36-37**,
   particle theory 18, 22
   primary colors 18
   wave theory 22
   wavelength interference 21

Walsh, Mike 14, **14**
Wavelength **see** Electromagnetic waves
White light 18, 45

X rays 25-27, **26,** 45
   instruments for detecting 39, **39**

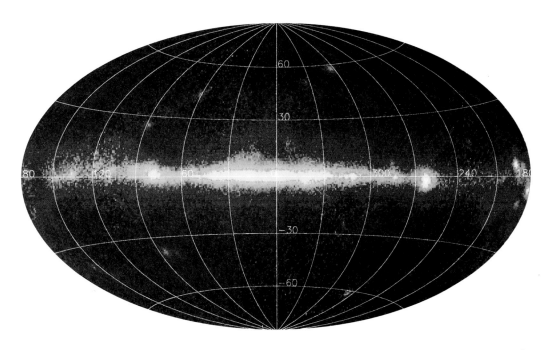

The high-energy gamma-ray sky

ILLUSTRATIONS CREDITS

Cover photograph, p. 2—Ian R. Howarth; Cover art—National Geographic Artist William H. Bond; p. 3, 8 top, 29—Davis Meltzer; pgs. 6-7—Peter Read Miller/*Sports Illustrated;* pgs. 8 bottom, 10 bottom, 11, 13, 16, 18, 19 top, 22, 23, 24, 26—Gloria Skurzynski; pgs. 9, 42-43—Roger H. Ressmeyer; p. 34 top—Roger H. Ressmeyer/Corbis; p. 10 top—National Geographic Photographer Jodi Cobb; p. 12—*Miami Herald;* p. 14—Chris Gotshall/Sea World of Florida; p. 15—Thermogram by Dr. Mike Walsh; pgs. 16-17—Frans Lanting; p. 19 bottom—Ralph Lee Hopkins; pgs. 20-21—Richard Faverty; p. 25—Ana Venegas; p. 27—Brown Brothers; p. 28—U.S. Air Force/Defense Nuclear Agency; pgs. 30, 47—Energetic Gamma Ray Experiment Telescope (EGRET) on the Compton Gamma Ray Observatory, courtesy Dr. Carl Fichtel and the EGRET Instrument Science Team; p. 31—Paul Chesley/Photographers Aspen; pgs. 32-33—Harold A. Weaver and T. E. Smith, Space Telescope Science Center, NASA; p. 34 bottom—Imke de Pater, et al, courtesy National Radio Astronomy Observatory/Associated Universities, Inc.; p. 35 top—Bernard J. Rauscher, University of Chicago and Center for Astrophysical Research in Antarctica; p. 35 bottom—Mark Hereld, Hien Nguyen, Bernard J. Rauscher and Scott A. Severson, University of Chicago and Center for Astrophysical Research in Antarctica; pgs. 36, 39 top—NASA; p. 37—Space Telescope Science Institute, NASA; p. 38 top—Goddard Space Flight Center and University of California, Berkeley; p. 38 bottom—Isabel Hawkins and Roger Malina, University of California/Berkeley Center for Astrophysics; G. Randall Gladstone, Dept. of Space Science, Southwest Research Institute; p. 39 bottom—J. Hunter Waite, Jr.; G. Randall Gladstone; W. S. Lewis, et al, Dept. of Space Science, Southwest Research Institute; p. 40—R. Williams, Space Telescope Science Institute, NASA; p. 41—J. Hester and P. Scowen, Arizona State University, NASA; Back Cover—Don King Films.

GLORIA SKURZYNSKI involves young readers in science and technology by showing how it affects their daily lives. Her nonfiction writing style has been widely praised. "Skillfully honed, comprehensive," said the Horn Book Magazine review of **Robots**. A starred Booklist review called **Get the Message** "Exceptionally readable and concise." A starred School Library Journal review for the author's cutting edge look at virtual reality, **Almost the Real Thing**, praised the "wonderful concept, covered with clarity and intelligence.... An excellent and lively book...." In 1992, the author won the American Institute of Physics Science Writing Award.

Gloria Skurzynski travels extensively across the country talking to students about writing, science, and technology. She says, "In the past dozen years we've learned more about the universe than in all the years since humans first began. It's an exciting time to be alive."

The author lives in Salt Lake City. Her husband, Ed, an aerospace engineer, and two of their daughters, both electrical engineers, are among the many experts she consults while researching a topic.

THE WORLD'S LARGEST NONPROFIT SCIENTIFIC AND EDUCATIONAL ORGANIZATION, THE NATIONAL GEOGRAPHIC SOCIETY WAS FOUNDED IN 1888 "FOR THE INCREASE AND DIFFUSION OF GEOGRAPHIC KNOWLEDGE." FULFILLING THIS MISSION, THE SOCIETY EDUCATES AND INSPIRES MILLIONS EVERYDAY THROUGH MAGAZINES, BOOKS, TELEVISION PROGRAMS, VIDEOS, MAPS AND ATLASES, RESEARCH GRANTS, THE NATIONAL GEOGRAPHY BEE, TEACHER WORKSHOPS, AND INNOVATIVE CLASSROOM MATERIALS.

VISIT OUR WEBSITE AT http://www.nationalgeographic.com OR GO NATIONAL GEOGRAPHIC ON COMPUSERVE.

THE SOCIETY IS SUPPORTED THROUGH MEMBERSHIP DUES AND INCOME FROM THE SALE OF ITS EDUCATIONAL PRODUCTS. CALL 1-800-NGS-LINE FOR MORE INFORMATION.